Relationship Skills Exercises

Marriages & Families:
Making Choices in a Diverse Society

TENTH EDITION

Mary Ann Lamanna
University of Nebraska, Omaha

Agnes Riedmann
California State University, Stanislaus

Prepared by

Lue K. Turner

Australia • Brazil • Japan • Korea • Mexico • Singapore • Spain • United Kingdom • United States

© 2009 Wadsworth, Cengage Learning

ALL RIGHTS RESERVED. No part of this work covered by the copyright herein may be reproduced, transmitted, stored, or used in any form or by any means graphic, electronic, or mechanical, including but not limited to photocopying, recording, scanning, digitizing, taping, Web distribution, information networks, or information storage and retrieval systems, except as permitted under Section 107 or 108 of the 1976 United States Copyright Act, without the prior written permission of the publisher.

For product information and technology assistance, contact us at
**Cengage Learning Customer & Sales Support,
1-800-354-9706**

For permission to use material from this text or product, submit all requests online at **www.cengage.com/permissions**
Further permissions questions can be emailed to **permissionrequest@cengage.com**

ISBN-13: 978-0-495-50597-6
ISBN-10: 0-495-50597-8

Wadsworth
10 Davis Drive
Belmont, CA 94002-3098
USA

Cengage Learning is a leading provider of customized learning solutions with office locations around the globe, including Singapore, the United Kingdom, Australia, Mexico, Brazil, and Japan. Locate your local office at: **international.cengage.com/region**

Cengage Learning products are represented in Canada by Nelson Education, Ltd.

For your course and learning solutions, visit **academic.cengage.com**

Purchase any of our products at your local college store or at our preferred online store **www.ichapters.com**

Printed in the United States of America
1 2 3 4 5 6 7 11 10 09 08

❧ TABLE OF CONTENTS ❧

Foreword

Chapter 1: How Many Bags Are You Checking?: What We Bring to Our Relationships From Our Families of Origin.....1
- Your State in the Union.....2
- Drawing Your Parents' Marriage.....12
- Genogram Activity.....13
- Life Events Timeline.....14

Chapter 2: Relationship Expectations.....15
- What Are Your Expectations About Relationships?.....16
- Deciphering Our Expectations: A Relationship Game.....21
- Putting it to Music.....22
- Keeping it Real: Remarriage and Stepparenting.....23

Chapter 3: What Are You Talking About?: Communication and Problem Solving.....27
- Lost in Translation: I-statements and Feeling Words.....28
- Who Has the Floor?.....30
- What the Squiggle?.....32
- Seeing Back to Back: Paying Attention to Nonverbal Cues.....33
- TV and Movie Clips.....34
- Steps Toward Negotiation.....35

Chapter 4: Sex and Intimacy.....39
- What's Your Sexual Script?.....40
- Where Are You Going?: Mapping Your Relationship Intimacies.....43
- Defining Intimacy.....45
- What's Your Comfort Level?.....46
- Giving and Receiving Feedback About Physical Touch.....47

Chapter 5: Financially Ever After.....49
- What Are Your Financial Values?.....50
- 10 Financial Principles.....52
- You Bought *What*?: Attitudes and Rules That Influence Our Finances.....54
- How Satisfied Are You With Your Financial Tasks?.....60
- How's Your Budget?.....62
- Doing Away With Debt.....65
- Financial Interview.....68

Chapter 6: Marriage Enrichment.....69
- Nurturing a Relationship Over Time.....70

❧ Foreword ❦

The overall purpose of this book is to enable individuals to gain a general understanding of relationships and the dynamic contributions each partner individually makes to their significant relationship. The intention is to have the reader explore their own habits, characteristics, qualities, and influences on decision making. These are things which ultimately affect who someone "is" in a relationship, and what she or he brings to a partnership. Another purpose is to provide a means for couples to explore their current relationship, be able to more fully understand themselves as well as their partner, and recognize the unique background and outlook each contributes to the relationship. This greater insight and understanding into the self and other can open up lines of communication and problem solving, which can lead to an enriched relationship.

These purposes will be accomplished by providing activities, assessments, worksheets, inventories, and discussion questions to enable users to evaluate themselves more closely in terms of behavioral patterns, habits, automatic thoughts, and family of origin and societal influences. These elements ultimately inform how an individual acts, reacts, and interacts with others. With new awareness and skills practicing, individuals and couples are better able to promote aspects related to relationship satisfaction, such as mutual understanding, consideration, and tolerance.

This book may be used by individuals or couples in a personal "at home" setting, in a small group setting (such as a marriage preparation or marriage enrichment class), or in a larger classroom setting as encouraged by an instructor.

❧ Chapter 1 ❧

HOW MANY BAGS ARE YOU CHECKING?
WHAT WE BRING TO OUR RELATIONSHIPS FROM OUR FAMILIES OF ORIGIN

YOUR STATE IN THE UNION

A large part of what we bring to relationships and how we act in relationships comes from our family of origin. Regardless of whether we want to do things similar to, or very different from our families, the first step is to recognize the environment of the family we came from in order to make more conscious and deliberate decisions about who we want to be in our own relationships or future families. In our families of origin we learn "rules" or ways of behaving with others that we often unconsciously carry into our future relationships with partners, friends, or roommates. Sometimes our frustration with, or dislike of others stems from their behavior violating some "rule" that seems intuitive to us, but others don't recognize it. Likewise, we often violate others' "rules" without even meaning to do so.

For example, maybe you have a boyfriend or girlfriend over for dinner to meet your family. Food is good, conversation is great, but the friend doesn't help clear the dishes from the table or wash them when the meal is done. Instead, the guest sits there while others do the work. Later, you and your family talk about what a nice personality the friend has, but she or he is rude and inconsiderate for not helping or even asking to help with dishes. Maybe this is not the kind of inconsiderate person to continue dating. She or he does not fit in the family. What does this dislike boil down to? The rule in your house is "Everyone helps with the dishes." The rule at the friend's house is "Mom always does dishes and I respect her domain in the kitchen." There are feelings of dislike and tension, and neither party understands or is even aware of what the "rules" are.

Fill out the "My Family of Origin Rules" inventory (pgs. 3-11) and then ask yourself the questions listed below. You may also have a partner fill out their own inventory and then discuss the answers to the questions together.

- What were the "rules" that governed your family of origin? What are some common themes that you notice?
- Which rules or themes do you still identify with or follow (whether you realized it or not)?
- Which rules or themes are you most comfortable with?
- What are some times in the past you have been upset with others because they didn't follow the "rules" (whether you or others were aware of them or not)?
- Which rules have most recently been called to your attention, or that you've noticed in your thinking or interactions with others?
- Which rules from your family of origin seem to honor family members and you would like to continue them? What will you do to do this?
- Which rules seem hurtful to family members that you do not want to continue? What will you do to do this?
- What are some "new" rules that you would want to follow in your current relationship with your partner, or follow in a future relationship?

MY FAMILY OF ORIGIN RULES
AN INVENTORY

Answer all of the following questions in reference to the family you grew up in. Your frame of reference for family may vary dependent on situations such as divorce, death, remarriage, or extended illness. Several related questions are grouped together. Use the lines to respond to the questions directly, or to write your own paragraphs about ideas what come to mind when reading the questions.

COMMUNICATION DYNAMICS

1. In the family you grew up in, who was allowed to talk to whom? Could children confront parents? How? Who was allowed to have disagreements? Were there any topics that were taboo to discuss? Who was or wasn't allowed to discuss them? How could family members approach someone they wanted to discuss something with? Could family members talk directly to one another, or was there sometimes a "go-between," or someone that carries a message from one person to another?

   _____
   _____
   _____
   _____
   _____
   _____
   _____
   _____
   _____
   _____
   _____
   _____
   _____
   _____
   _____
   _____
   _____

2. How were disagreements or arguments handled in the family? Between spouses? Between parents and children? Between children?

   _____
   _____
   _____
   _____
   _____

3. What were ways love and affection were expressed in the home? In public? Between spouses? Between parents and children? Between siblings? Who was allowed to express love or affection? Were others outside the family allowed to see expressions of love or affection between family members? Was love and affection expressed verbally or nonverbally? Did gender make a difference in terms of expression?

   _____
   _____
   _____
   _____
   _____
   _____
   _____
   _____
   _____
   _____
   _____
   _____
   _____
   _____
   _____
   _____
   _____
   _____
   _____
   _____

4. What were acceptable ways for a family member to express anger or hurt? How did others react to these feelings?

   _____
   _____
   _____
   _____
   _____
   _____
   _____
   _____
   _____

CONFORMITY

5. How different were family members from one another (clothing, hairstyle, language, hobbies, gender roles, personality)? How were differences viewed? How was someone allowed to be unique in your family? What were acceptable ways to express oneself? What would happen in your family when someone expressed their individuality? How was "time alone" or "time to oneself" or "time with one's friends" viewed?

_____

BOUNDARIES

6. How open was your family to new influences and experiences? Who and how often were others allowed to come inside the house, including friends, neighbors, extended family, and even strangers? How were visitors handled (drop-in or stop-by, certain hours only, must call ahead, just walk in, keys given out to others)? Who could visit and how long could they stay? Did it make a difference if the visitor was family vs. a friend?

7. How was privacy handled in the home between family members? Who could have it and how was it achieved? How would you rate the level of privacy in your home? How did you know someone wanted privacy? What did a closed door (to a bedroom, bathroom, or any room) really mean? Were there any rooms certain members of the family were not allowed access to? Which, if any, doors had locks and how were they used? Were secrets kept in your family? What were they and who was allowed to know? Was your family one that kept family issues to themselves, or did they invite others (friends, other family) to know?

HOLIDAYS

8. Which holidays did your family celebrate and why? What would your family do for these holidays? Who was allowed to participate in celebrations? Which traditions mean the most to you? What are some traditions your family has which have not changed over the years?

SEXUALITY AND MODESTY

9. How was the topic of sexuality handled in your family? How could one express their sexuality? What was acceptable? Who was allowed to talk about sex or sexuality? To whom? When? How were issues of modesty (in regards to dress, or covering the body, or vulgar sexual language) addressed in your home?

_____
_____
_____
_____
_____
_____
_____
_____
_____
_____
_____

VIOLENCE AND ABUSE

10. What types of violence or inappropriate acts (physical violence, sexual abuse, emotional or verbal abuse) were you aware of, or suspect in your home? How safe did members of the family feel in the home and in the presence of other family members? How appropriate were relationships between spouses? Between parents and children? Between siblings? If violence of abuse was going on between family members, who was allowed to know in the family? Outside the family?

_____
_____
_____
_____
_____
_____
_____
_____
_____
_____
_____
_____
_____
_____
_____
_____

HOUSEHOLD WORK AND GENDER ROLES

11. What types of work or jobs were performed on a routine basis inside or outside the home? Who did what types of jobs? How was the work of the house divided up among family members? What did parents do? What did children do? What types of work were hired out for? Why was it divided the way it was? How did family members feel about the arrangement? What did it mean in your house to be a female? To be a male? Were female and male family members seen as equal, or treated equally in your home? How do you know?

RELIGION AND SPIRITUALITY

12. How was religion or spirituality handled or incorporated in the family? Who participated in religious activities and why? What were some religious or spiritual beliefs that were followed? Were they followed by personal choice or was participation enforced? Was spirituality viewed as an individual or family endeavor or both? How did religion or spirituality influence family activities or decisions made? What traditions, practices, or customs were regularly practiced that were based in religion or spirituality? Was religion or spirituality experienced as one part of life, or was it interwoven through all parts of life?

DRAWING YOUR PARENTS' MARRIAGE

Often we learn how to "do" marriage from others. Many times our marriage frame of reference is learned from our parents and what we saw, or knew, about their marriage. Exploring your feelings about your parents' marriage and what you learned from that relationship can help you make more deliberate decisions about what you want in your own marriage and partnership.

Using large pieces of paper (butcher paper) and markers, draw a pictorial representation of your parents' marriage. To get started, fill in the following sentence fragments:

My parents' marriage is (or was) like a _____.
OR
My parents' marriage reminds me of a _____.

Examples:
My parents' marriage is like a garden where everything gets fertilized and grows.
My parents' marriage reminds me of a deadly train wreck.

Draw, fill in, and color your picture any way you desire based on your feelings or observance of your parents' marriage. It's OK to be creative, politically incorrect, or symbolic.

You may also have your partner do a drawing and then take time to explain your drawings to each other. Ask yourself or discuss with your partner the following questions:

➢ Why did you choose the analogy you did to represent your parents' marriage?
➢ What did you like about your parents' marriage? What did you not like?
➢ How does your experience of your parents' marriage influence your ideas about marriage or how marriage should be?
➢ How does your experience of your parents' marriage inform what types of qualities you are seeking in a partner? How does it inform qualities you are trying to cultivate for yourself?

You may also consider drawing a representative picture of what your ideal marriage would be like. What would you do to work toward achieving this?

GENOGRAM ACTIVITY
A HOMEWORK ACTIVITY

A good way to get to know a partner better is to make a genogram, or a family tree, of their family. Using circles to represent females, squares to represent males, and lines to demonstrate connections or relationships between people, interview a partner about their family. You may also use color to code or categorize things, or create a key to any symbols used for the genogram. Be sure to ask about or find out the following:

- ✓ Ages of family members
- ✓ Significant family deaths or diseases
- ✓ Dates of marriage or divorces
- ✓ Relationship dynamics between family members (between siblings, spouses, parents and children)
    - ➢ Who do you trust?
    - ➢ Who is most supportive?
    - ➢ Who often gets excluded?
    - ➢ Who is their tension between?
    - ➢ Are there any "favorites" in the family?
- ✓ Where (parts of the country or world) family members are or where they came from
- ✓ Types of jobs held by family members

If you aren't able to interview a partner, you may make a genogram depicting your own family.

- ▶ What did you learn about your partner's family or your own family that you didn't realize before?
- ▶ Based on the genogram, what types of family patterns have become apparent? How do these affect you or your relationship?

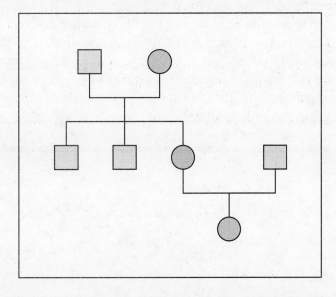

13

LIFE EVENTS TIMELINE
A HOMEWORK ACTIVITY

Using large sheets of paper (butcher paper), make a detailed timeline of your life. Even the smallest events or activities in our lives can still have an influence on who we are and how we relate to others. Anything can be included on the timeline, but be sure to include the following:

- ✓ Births and deaths of family members, especially during your lifetime
- ✓ Major and minor milestones (staring kindergarten, moving away from home, knee surgery, etc.)
- ✓ School, religious, or community activities in which you've participated
- ✓ Dates related to significant relationships you've had (first date, breakups, engagements, relationships with previous partners)
- ✓ Dates of significant family events (parents' divorce, military service, big vacations)
- ✓ Dates of meeting role models, mentors, or other influential people in your life
- ✓ Anything else that would help provide a snapshot picture of your life and what makes you who you are

Instead of drawing your own timeline, you can also work with a partner taking turns interviewing and drawing each other's timelines.

- ➢ As you look at your timeline, what are the parts that stick out or seem significant to you? Are there any themes or patterns that are apparent?
- ➢ Does anything seem missing from your timeline that you think was influential in your life?
- ➢ How has the time spent in past relationships influenced who you are today? What did you learn about yourself during that time?
- ➢ Based on family or previous partner relationships, what have you learned from your experiences about how to "do" relationships?
- ➢ What childhood events influence you and your relationships today? (examples: your family declared bankruptcy, a good friend or classmate died, your parents always attended your school activities)
- ➢ Overall, what did you learn or realize about yourself from drawing your timeline?
- ➢ If you drew your partner's timeline, what did you learn about your partner and their personal history that you didn't know before?

Chapter 2

RELATIONSHIP EXPECTATIONS

WHAT ARE YOUR EXPECTATIONS ABOUT RELATIONSHIPS?

A large portion of happiness in marriage and relationships is based in expectations. The more our relationship expectations are met, the greater likelihood we will be satisfied with the relationship. For this reason, it is essential to identify and examine our own relationship expectations, as well as our partner's, or ideal partner's, expectations about the relationship. Once we recognize our expectations, they can be evaluated for whether they are realistic, feasible, or helpful to the relationship. It is beneficial to change expectations that are not realistic, and to discuss relationship expectations with our partner.

Fill out the "What Are Your Relationship Expectations?" inventory (p. 17), rating to what degree you identify with each statement. You may fill this inventory out on your own, as well as have a partner fill out their own copy. After you have completed the inventory, look at the "Relationship Expectations Evaluation" exercise (p. 18). This exercise will describe each statement from the inventory in more detail and enable you to evaluate your responses. Fill out "What Are Your Relationship Expectations?" inventory first before looking at the "Relationship Expectations Evaluation" exercise.

WHAT ARE YOUR RELATIONSHIP EXPECTATIONS?
AN INVENTORY

Circle your response to the following statements to indicate to what degree the statement reflects an expectation you have of a romantic relationship.

	Strongly Disagree		Undecided		Strongly Agree
1. If my partner loves me, she/he would *instinctively* know what I want and need in order to be happy.	1	2	3	4	5
2. No matter how I behave, my partner should love me simply because she/he is my partner.	1	2	3	4	5
3. I can change my partner by pointing out her/his inadequacies, errors, and other flaws.	1	2	3	4	5
4. Either my partner loves me or doesn't love me; nothing I do will affect the way she/he feels about me.	1	2	3	4	5
5. The more my partner discloses positive and negative information to me, the closer I will feel to her/him and the greater our marital satisfaction will be.	1	2	3	4	5
6. I must first feel better about my partner before I can change my behavior toward her/him.	1	2	3	4	5
7. Maintaining romantic love is the key to marital happiness over the life span for most couples.	1	2	3	4	5
8. Marriage should always be 50/50 partnership.	1	2	3	4	5
9. Marriage can fulfill all of my needs.	1	2	3	4	5
10. Couples should keep their problems to themselves and solve them alone.	1	2	3	4	5

From Larson, J. H. (Winter, 2006). Overcoming myths about marriage. *Marriage & Families*. Provo, UT: School of Family Life, Brigham Young University, p. 6.

RELATIONSHIP EXPECTATIONS EVALUATION

All of the expectations listed in the "What Are Your Relationship Expectations?" inventory are actually myths. The following is a way to assess your degree of belief in these myths:

    High: If you agreed or strongly agreed with 5 or more myths
    Moderate: If you agreed or strongly agreed with 3 or 4 myths
    Low: If you agreed or strongly agreed with 2 myths or less

Repeated below are the marriage myths along with the reality statement that counters it.

- **Myth 1** If my partner loves me, she/he would *instinctively* know what I want and need in order to be happy.

    Reality Check: If my spouse really loves me, she/he will openly and respectfully tell me what she/he needs and not expect me to read her/his mind.

- **Myth 2** No matter how I behave, my partner should love me simply because she/he is my partner.

    Reality Check: Your spouse will love you to the extent that you are loveable, and that's based largely on your behavior.

- **Myth 3** I can change my partner by pointing out her/his inadequacies, errors, and other flaws.

    Reality Check: I can positively influence my spouse's behavior if I know how, and that can be learned. But nagging does not work.

- **Myth 4** Either my partner loves me or doesn't love me; nothing I do will affect the way she/he feels about me.

    Reality Check: If I behave more lovingly, she/he will love me more.

- **Myth 5** The more my partner discloses positive and negative information to me, the closer I will feel to her/him and the greater our marital satisfaction will be.

    Reality Check: The expression of positive thoughts and feelings increases marital satisfaction the most. If you have something negative to disclose, watch how you do it so as not to offend.

⊘ **Myth 6**   I must first feel better about my partner before I can change my behavior toward her/him.

Reality Check: Part of being married is learning that you sometimes have to do things for your partner that you would rather not do, simply to please your partner. As she/he becomes happier, she/he will likely reciprocate with pleasing behaviors too, and you too will be happier (your feelings change). Plus, you will feel much better about yourself as a result of changing your behavior first without hesitating too long.

⊘ **Myth 7**   Maintaining romantic love is the key to marital happiness over the life span for most couples.

Reality Check: It takes compassionate and altruistic love, too, to preserve your marriage.

⊘ **Myth 8**   Marriage should always be 50/50 partnership.

Reality Check: Your marriage will be stronger if you focus on pleasing your partner and making sure you are doing all you reasonably can to contribute without keeping a tally.

⊘ **Myth 9**   Marriage can fulfill all of my needs.

Reality Check: Marriage can fulfill many of your needs, and the others can be fulfilled by other appropriate people. For example, if your spouse doesn't enjoy playing golf, play golf with a friend instead to fulfill that need of "companionship on the golf course."

⊘ **Myth 10**   Couples should keep their problems to themselves and solve them alone.

Reality Check: Keeping your problems quiet and going it alone often leads to failure. Get trusted others or a neutral third party to help you.

From Larson, J. H. (Winter, 2006). Overcoming myths about marriage. *Marriage & Families.* Provo, UT: School of Family Life, Brigham Young University, p. 7-10.

RELATIONSHIP EXPECTATIONS REVIEW

By yourself or with a partner contemplate or discuss the following questions in response to the "What Are Your Relationship Expectations?" inventory and the "Relationship Expectations Evaluation."

- ➤ What was your reaction to your score on the "What Are Your Relationship Expectations?" inventory?
- ➤ For the myths that you believed to be true, or you strongly agreed with them, what evidence were you basing your decision on? How credible is your evidence? Have you ever encountered other evidence that contradicts this myth?
- ➤ What "mythical" behaviors do you exercise that have fed your past relationships? Your current relationships?
- ➤ Knowing what you know now about relationship myths, what personal beliefs pertaining to relationships would you be willing to re-examine or change?
- ➤ Looking at the list of Reality Check statements, what new behaviors and ways of thinking would you need to adopt in order to incorporate that "reality" in your current or future relationships?

DECIPHERING OUR EXPECTATIONS:
A RELATIONSHIP GAME

Are you sure of what your partner's relationship expectations are? Do you understand what expectation drives some of their thoughts and actions? This activity can be done as a couple, or with several couples together.

The object of the game is for couples to learn more about each other's personal preferences and expectations of the relationship. Beforehand, couples can submit questions or scenarios to answer, or a third partner can come up with questions. Questions can be on a wide range of topics, but should focus on individual behaviors or expectations of the relationship. Some examples of questions could be:
- Given the choice between watching the "big game" or attending a good friend's wedding with you, what would they choose?
- Who do you expect will monitor the finances in your relationship?
- How does your partner like to be comforted after a bad day?
- If your partner is unusually quiet, what does that mean and how do they expect you'll react?
- What would your partner rather receive for Valentine's Day: a gift, a hug, or stimulating conversation?

For the game, one partner from each couple should leave the room so the questions being asked are not heard. The moderator will ask each remaining partner the same question (example: How many hugs a day does your partner desire?) and each partner will write their answer on an index card or poster. After answering 4-5 questions, the missing partners will re-enter the room. The moderator will then ask each question again so that the partner who was out of the room can answer. The index card or poster with the answer from the other partner can be held up for comparison after an answer is given. Couples or a moderator can keep track of how many questions they got correct, as they were accurately able to know their partner's behavior or expectation.

In doing this activity as a couple (not as a group), 8-10 questions can be thought of by each partner and put in two different bowls. Take turns with each partner pulling a question from their own bowl, reading it aloud, and guessing what their partner would answer. Then have the other partner respond to the question and proposed answer. Continue to take turns until all questions are answered.

At the conclusion of the game discuss with your partner the following:
- ✓ How well does each of you understand the expectations your partner has of the relationship?
- ✓ What did you learn about your partner that you didn't know before? How do you feel about this new information?
- ✓ Where do you go from here? What relationship expectations need to be addressed or talked about, or compromised on?
- ✓ What did you learn about your own expectations in the relationship?
- ✓ What questions do you have for your partner that you still need or want to ask?

PUTTING IT TO MUSIC
A HOMEWORK ACTIVITY

Many times it is not easy to communicate what our expectations are of our partner or the relationship in general. The object of this activity is to share with a partner your expectations of love and relationship behaviors through music. Each partner can find a song or organize a collection of songs that to them expresses a definition of love, how a relationship should be, or what she/he is looking for in a relationship. Several songs or parts of songs may be needed to fully communicate this. Arrange a time where you can play your songs for each other, or else swap your recording(s), listen on your own, and arrange a time to meet together later. This is a fun activity designed to help couples realize the expectations they have for a relationship and to talk about them.

This activity can also be done on your own. Collect songs that you feel talk about or represent what you would want in an ideal relationship.

Assess for yourself or discuss the following with a partner:

- ✓ Based on your choice of music, what do you realize about your own expectations regarding relationships?
- ✓ What does the music you chose communicate about love, definitions of romance, and commitment to the relationship? What does the picture look like that your music portrays?
- ✓ Do you think your expectations are realistic? Do you think the songs you chose promote ideas of love based in fantasy or reality?
- ✓ What did you feel about your partner's song selections?
- ✓ Based on your music selections, what areas of overlap do you and your partner have in terms of relationship expectations? Expectations of a partner?
- ✓ In terms of expectations of a relationship or a partner, what do you and your partner differ on?
- ✓ Which expectations do one or the other, or both of you, feel you need to adjust?

KEEPING IT REAL:
REMARRIAGE AND STEPPARENTING
A HOMEWORK ACTIVITY

The reality today is that a significant number of marriages include at least one spouse who has been married before. Related to this, there are also opportunities to create stepfamilies and stepparents through marriage. What happens to our expectations of love and relationships with a remarriage? What are realistic expectations to have when becoming a stepparent at the same time?

There are two parts to this activity, one focused on remarriage (p. 24) and one focused on stepparenting (p. 25). Each of these is a complex family situation that requires a conscious decision to be involved and adjust expectations. Two ways to prepare for a remarriage or to be a stepparent is to both 1) analyze for yourself your own feelings about the situation, and 2) learn from others to anticipate what some realistic expectations would be. Although the activities here apply one step per situation, both could be carried out for each scenario.

REMARRIAGE SELF-ASSESSMENT

Conduct a self-interview and analysis of your own feelings about remarriages and entering into a remarriage. Answer for yourself the following questions to assess your own openness or expectations of a remarriage:
- If you have never been married, would you consider marrying someone who already had? Why?
- If you have preciously been married, are you looking for someone who has been through a similar experience or someone who has not been married before?
- What are your assumptions about individuals who have already been married at least once before and are now single?
- Would it matter to you if you were someone's second spouse? Third? Fourth?
- Does it make a difference to you if the partner you are considering was previously divorced versus widowed?
- What beliefs do you have about how remarriages are different from first time marriages?
- What would you expect to be different in a relationship between two people in which at least one partner has been married before, versus a relationship in which neither person had been married before?
- What real-life examples have you witnessed of remarriage? What was your impression? What have you learned from this?
- What media or TV examples have you witnessed of remarriage? What was your impression? What have you learned from this?
- In contemplating remarriage, what role does your age play in your decision?
- In contemplating remarriage, what role do children (either theirs or yours) play in your decision?

After you have answered these questions ask yourself the following:
- Based on your answers, what did you learn about yourself?
- Did any of your answers surprise you?
- What issues related to remarriage had you not thought of before?
- Overall, would you consider entering a remarriage?

STEPPARENTING INTERVIEW

Choose at least one person you know that became a stepparent through marriage. It would be best to interview one stepparent who is male and one who is female. Set aside a time to interview them both either in person or on the phone. Prepare several questions ahead of time that you would like to explore related to their marriage, stepparenting, and transitions as a stepfamily. Even if you are not anticipating this as a life situation for yourself, experience this approach as a way to gather crucial information to make that decision. Listed below are some sample questions that may be included in the interview. Be sure to ask appropriate follow-up questions and any other questions that seem pertinent.

Possible questions to ask:
- How did you become a stepparent? Was this something you ever planned on? How long have you been a stepparent?
- How did the age of the child (or children) factor into your decision to be a stepparent?
- What do you do as a stepparent? How do you see your role? Is your role now different than what you thought it would be originally?
- What do the stepchildren call you, or refer to you as? How was that decided?
- What are some of the good things about being a stepparent? Did you anticipate these ahead of time?
- What are some of the challenging things about being a stepparent? Did you anticipate these ahead of time?
- Is your relationship with your spouse what you expected it would be? Why?
- How do you handle couple time or alone time with your spouse?
- What role does the previous spouse play in your lives as a family? Did you expect this?
- What did you expect stepparenting would be like? Where did you learn those expectations from? How does actually being a stepparent compare to what you expected?
- What advice would you give to others who will become stepparents?

After the interview discuss or write down your impressions of the interview.
- What did you learn about stepparenting that you hadn't previously known?
- Which responses confirmed beliefs you already had about stepparenting?
- Which responses challenged beliefs you have about stepparenting?
- How do you see a marriage with stepparenting as different from a marriage where no children are involved?
- Would you consider being a stepparent? Why?

❧ **Chapter 3** ❧

WHAT ARE YOU TALKING ABOUT?
COMMUNICATION AND PROBLEM SOLVING

LOST IN TRANSLATION: I-STATEMENTS AND FEELING WORDS

Many times when we express our feelings to others, we use words and phrasing which actually communicates that we are putting blame for our feelings on another person. This can raise defenses and contribute to the "heat" often felt in arguments with our partner, family members, or friends. The purpose of using I-statements is to express to others what we are feeling in a way that doesn't put anyone on the defensive. An I-statement would express the underlying feeling, as well as describe the action or event that precipitated the feeling. For example, an I-statement could be:

> *"I feel so upset and disorganized* **[the feeling]** *when the kitchen counters aren't clean when I get home from work".* **[the event]**

Practice forming I-statements by using the "I-statements" worksheet (p. 29). As you encounter conversations with your partner, friends, or family members, practice phrasing your feelings in terms of I-statements.

I-STATEMENTS WORKSHEET

Practice rephrasing the following statements into I-statements that reflect ownership of a feeling. Identifying an action of what precipitated the feeling is important, and can be fabricated for these examples. As you practice writing these I-statements, really tap into feeling words that are descriptive. Don't just use "angry" or "mad;" try to use words that are more descriptive (e.g. frustrated, dismissed, unimportant, left out, forgotten, hurt, etc.)

Example: "I hate it when you're an idiot!"
I-statement: "I feel worried [feeling] about paying our bills on time when you forget to cash your pay check." [event or action]

1. I *loved* (said sarcastically) the empty gas tank in my car this morning! _____

2. You can be such a jerk. _____

3. Why did you buy that? Do you think I'm made of money? _____

4. I shouldn't have to *remind* you to change the baby's diaper. _____

5. You make me so happy. (This is a positive feeling, but can be rephrased to explain specifics and show ownership of feelings.) _____

6. We always watch what *you* want. Does my opinion even matter? _____

7. Leave me alone! All I want to do is relax after work. _____

8. _____ (Fill in your favorite complaint statement here and then change it to an I-statement.) _____

WHO HAS THE FLOOR?

Oftentimes couples over-talk or interrupt each other during discussion, especially if the topic revolves around a problem or a somewhat heated issue. Often, we are already thinking of our "come-back" argument and fail to listen to what our partner is saying. Something that can help during the discussion is to slow down the communication and expression process and to take turns talking and understanding each other.

With your partner or a friend, practice I-statements and reflective listening during a conversation. To do this use a piece of flooring, like a 12 x12 inch vinyl square or a piece or remnant carpet that can be easily passed back and forth between you and your partner. The ground rule for the conversation is:

- Only one person at a time talks—this is whoever has the "floor" (vinyl or carpet piece).

For this exercise do the following:

1. The person speaking with the floor sensitively shares 1-2 concerns or complaints (no dumping 20 different issues while you have the floor). Use I-statements as you share your concerns and feelings.

2. Pass the floor to the other person. The listener needs to reflect back the feeling or concern (to demonstrate listening) before responding to it with additional I-statements or ideas.

3. Pass the "floor" back and forth between each other as you take turns being the listener or the speaker. The goal is to promote understanding, demonstrate listening skills, and share feelings in a non-accusatory way.

4. The exercise can be concluded upon mutual agreement, solving of an issue, or a gaining of common understanding. If things become too heated or emotional during the discussion, a time-out can be taken with an agreement to finish the discussion later (meeting again in 10-30 minutes).

The exchange and passing of the "floor" can go on several times with each person taking turns during the conversation being the speaker and listener. As a speaker, be sure to keep your comments brief to allow for accurate reflection by the listener when the floor is passed to them (a person can only remember so much). As you and your partner get better at listening to understand, and using I-statements, the use of the actual "floor" will become less necessary.

At the conclusion of the exercise process as a couple, or ponder the following for yourself:

- ✓ Was it hard to not interrupt if it wasn't your turn to have the "floor?"
- ✓ How intently were you listening to try and understand your partner? Were you really listening or were you busy thinking of your next argument or comeback?
- ✓ How accurately were you able to reflect your partner's feelings or concerns back to them to demonstrate understanding?
- ✓ How were you at giving up the "floor" to someone else? Were you able to give someone else a chance to speak uninterrupted?
- ✓ What did you learn about your communication skills during this exercise? What do you do well? What are skills to improve upon?

WHAT THE SQUIGGLE?

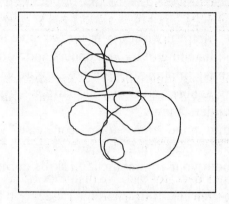

Are you able to find the words you need to communicate to your partner or others? Do you find yourself being understood? How well do you listen to what your partner or others are trying to tell you? Practice your listening and communication skills with this Squiggle Drawing exercise.

Work with your partner or with a friend. One person will need to be the Listener and one will need to be the Describer. The Listener needs a blank piece of paper and pen and needs to be blindfolded. The Describer's task is to describe a Squiggle to their partner for them to draw. You may time the activity (3-5 minutes) or just continue until the drawing is complete. After a turn, switch roles with your partner. Have the Describer rotate the Squiggle drawing 1-3 times, or pick a different place to start, and give instructions on how to draw it from that point of view.

How did each of you do? Remember, regardless of how the picture turns out, the overall focus of the activity is on listening and descriptive communication. Write down or discuss the following with your partner:
- ✓ Did you prefer to be the Listener or Describer? Why? How did each of the roles feel to you?
- ✓ Which role (Listener or Describer) was easier for you to take on? Why?
- ✓ As the Describer, how did you handle it if your partner wasn't drawing what you felt you were describing?
- ✓ As the Listener, did you feel you understood what your partner wanted from you? How did you know that you were, or were not, on track?
- ✓ As the Listener, did you ask clarifying questions of your partner if you weren't sure you understood? Why? What either encouraged or discouraged you to ask important and essential questions?
- ✓ Did each of you view and describe the Squiggle in the same way? Was there more than one way to describe the Squiggle to another person? Can two people see the same thing and describe it differently?
- ✓ What are other areas in your relationship where perhaps you want the same outcome, but you view the process of getting there differently? What can you do to work together in a more effective way to achieve your goals?

SEEING BACK TO BACK:
PAYING ATTENTION TO NONVERBAL CUES
A HOMEWORK ACTIVITY

How well do you know and understand your partner's nonverbal cues, such as voice inflection and intonation, and subtle body cues? This exercise can be done with a partner or a friend, but is easier with someone you know well.

Sit on the floor with your partner back-to-back, so that your backs are leaning up against each other. Each partner should put their hands down at their sides and hold the other partner's hands in a comfortable manner. Each person can then take turns just talking to each other about a concern, a worry, or just how their day was. The purpose of the exercise is to practice good listening skills as well as to pay attention to subtle nonverbal cues such as voice inflection and intonation, and small body cues from the hands and back. After about 10 minutes, when each partner has been able to have a turn expressing ideas and listening, discuss or write down the following:

- Was it more difficult to understand or listen to someone when you weren't able to see them, especially their facial expressions? Why? What are some situations where it would be crucial to understand your partner, but you wouldn't be in a position to read their face?
- What was it like to listen to someone you could touch, but not see? What did you notice about the inflection, intonation, or quality of your partner's voice as they spoke? Were you able to decipher your partner's emotions or meaning even though you couldn't see your partner?
- How did you do with communicating your own meaning without using facial expressions or your hands? Were you able to get your point across with just your voice and subtle cues?
- What nonverbal signals were you getting from your partner's hands? Did they feel warm, cold, tense, relaxed, or sweaty? What did you notice about their breathing? Did anything about your partner's hands or breathing change over the 10 minutes? What meaning did you take from this?
- What do you tend to pay more attention to, the actual verbal, or the subtle nonverbal communication that accompanies it?
- What can you interpret from this exercise in terms of overall communication in your relationship?

TV AND MOVIE CLIPS
A HOMEWORK ACTIVITY

Sometimes we receive our ideas of how couples should communicate from TV and movies. Have yourself and your partner each record a clip from a favorite TV show or sitcom, or identify a clip from a movie, that depicts a couple communicating in some way. It can be a clip of how a couple handles a disagreement, or how they express love or affection to each other. Pick a time where each of you can show your clips to each other. As part of your sharing, answer the following questions and discuss with your partner:

- Why did you choose the TV show or movie clip that you did?
- What was significant about this couple that stuck out to you?
- Evaluate the onscreen couple's communication skills based on what you know about good communication. How did they do?
- What did the onscreen couple do to communicate that is similar to you and your partner? What is dissimilar?
- What behaviors, attitudes, or statements did the onscreen couple express that you would like to incorporate in your relationship? What were positive, affirming things they said or did to build their relationship? What kind of toxic things did they express that you feel you need to avoid or eliminate in your relationship?

This activity can also be completed on your own. Analyze for yourself how the clip informs your views on couple communication. What would you want to experience in your ideal relationship?

STEPS TOWARD NEGOTIATION
A HOMEWORK ACTIVITY

Negotiation or compromise is a natural part of any long term relationship. The following is an exercise designed to take you and your partner, or you and a friend, through some simple steps of negotiation to help reach agreement or compromise on important issues.

This process can be used with any type of relationship where agreement on an issue is difficult or previous discussions have lead to aggravation or upset feelings. Listed below are steps to follow in negotiation as well as an example that follows the steps.

1. Define the issue or problem to solve.

   Example: Too much of family budget is going toward eating out

2. Use I-statements and reflective listening to truly understand how the other person feels or where they are coming from on this issue. This step can be carried out verbally, and the main concerns can be written out in the form of I-statements. It is important that each person feels their position is understood. If one or both people feel discounted or that their opinion is unimportant, negotiations will most likely break down, since at least one person does not feel respected or understood. Feeling validated and understood is essential to this process where compromise is sought.

   Example:

   Partner #1 – "I worry about wasting money on eating out. I'm concerned the children aren't getting proper nutrition."

   Partner #2 – "The kids are tired and hungry and I never have ideas on what to cook or don't have ingredients on hand. Going out to eat feeds the kids quickly."

3. Brainstorm possible solutions. Don't "edit" or evaluate any solutions or ideas, just brainstorm. The first solution can even be ridiculous. It lightens the mood and after a crazy suggestion, any other alternative could seem better.

   Example Solutions:

   - have the kids skip lunch (this is the ridiculous one)
   - allocate more of the budget to eating out and food; spend less money on other things (family vacations or toys for the kids)
   - eat every meal at home
   - eat out once a week
   - grocery shop more often so ingredients are in the house
   - buy a new cookbook with interesting recipes to try
   - plan menus so everyone will know what will be eaten for every meal

4. Evaluate the pros and cons of each solution. List pros and cons by each solution.

   Example:

   - Eat every meal at the home   + save money
   + opportunities for better nutrition
   - no flexibility for meals out
   - not always have time to cook at home

   - Grocery shop more often   + ingredients in house to cook with
   - takes planning ahead

5. After the pros and cons evaluation, each partner then circles solutions they would consider all or in part, or be open to.

6. Look for overlapping circles where both partners have circled the same thing—these are solutions that both of you would be willing to consider, compromise on, or talk about more. During this stage often new ideas or solutions can be discussed. Also, sometimes solutions from the list are combined or modified, or more than one is selected to try.

   Example Solution:

   - Grocery shop more often
   - Menu/plan meals a week at a time
   - Eat out once a week

   Plan: As a family we will discuss on Sunday nights what we want to eat for breakfasts, lunches, and dinners for the entire next week. We will write the entire menu on a calendar for the refrigerator. Based on the menus, we will grocery shop on Mondays for all items needed the next week for the planned meals. Only items on the grocery list pertaining to the planned meals will be purchased. We will also pick one night during that week when we will go out to eat as a family.

7. Agree on a plan of action to solve the problem and try this for a limited amount of time (e.g. 2 weeks).

8. Re-evaluate. How is it working? What needs to be tweaked? At this point you can re-evaluate how things are working and make any necessary changes to your plan of action.

STEPS TOWARD NEGOTIATION WORKSHEET

Use the following worksheet to map your own negotiation steps with a partner or friend.

1. Define the issue or problem to solve. Be specific: _____
   _____
   _____

2. Write down the key concerns from each person in the form of I-statements. Make sure each person feels their point of view is understood.

3. Brainstorm possible solutions to the problem. Remember, do not edit or evaluate any potential solution, just write it down. You can offer a ridiculous solution, too.

4. Evaluate the pros and cons of each solution proposed. You can write them in next to the solutions above using + or – symbols, and writing each point in the margin.

5. Have each person circle solutions they like or would consider all or in part. Rewrite the solutions here where some common ground was found:

6. Select a solution to try from the list. Remember this is when solutions can be combined or modified, or even more than one solution can be implemented. Exact details as to how to carry out the solution can be worked out now. Write your selected solution here along with the details as to how this will be carried out (who will do what, when, and how):

_____
_____
_____
_____
_____
_____
_____
_____

7. Decide how long to implement the solution before evaluating its success. This re-evaluation time can be anywhere from a few days to a few weeks. Write down a date when both people can discuss how the new solution is working: _____
_____

8. At the re-evaluation, discuss the following:

   - What about the solution is working? _____
   _____
   _____

   - What about the solution isn't working? _____
   _____
   _____

   - What outcomes are happening that we didn't anticipate with this solution?
   _____
   _____

   - What needs to be changed or tweaked to make the solution better? Do we want to stay with this one, or do we want to try something else from the solution list?
   _____
   _____
   _____
   _____

… **Chapter 4** …

SEX AND INTIMACY

WHAT'S YOUR SEXUAL SCRIPT?

How we think and feel about aspects of sexuality is in part determined by the messages we receive from our family, friends, media, and society across our lifespan. Answer the questions on the "What's Your Sexual Script?" inventory (pgs. 41-42) to determine your cultural learning about sex and what informs your attitudes and current mindset.

After you have completed the inventory, look at your answers.

- How would you characterize your sexual script?
- What types of patterns do you notice?
- Who, or what, has primarily influenced your sexual script?
- What aspects of your sexual script surprised you?
- Are there any aspects of attitudes of your sexual script that you are uncomfortable with, or want to change?
- How has your sexual script influenced you in previous relationships? How do you see it influencing your current or future relationships?

Have a partner complete the "What's Your Sexual Script?" inventory on their own. Come together to discuss and talk about your responses and various influences you each have had that inform your ideas and attitudes about sex.

- What were the similarities?
- What areas of difference need to be addressed?
- What did you learn of your partner's attitudes and beliefs about sex that you did not realize before?

WHAT'S YOUR SEXUAL SCRIPT?
AN INVENTORY

How we think and feel about aspects of sexuality is in part determined by the messages we receive from our family, friends, media, and society across our lifespan. Answer the following questions to help determine your cultural learning about sex and what informs your attitudes and current mind set.

1. How do you define sex? _____
   _____

2. What is the purpose of sex? _____
   _____

3. How did you first learn about sex? How would you characterize that experience? What, if anything, would you change about how you first learned about sex? ____
   _____
   _____
   _____

4. How much was sex or issues regarding sexuality talked about in the home you grew up in? How was the topic of sex handled? _____
   _____
   _____

5. What type of role did parents, friends, and media (movies, TV, music, Internet) play in your own education regarding sex? _____
   _____
   _____
   _____

6. What types of sexual activities or behavior are acceptable to you? _____
   _____
   _____

7. How frequently should a committed couple engage in sexual activity? _____
   _____

8. How much time should any single sexual experience with a partner take? _____
   _____

9. Who is responsible for initiating sexual activity? _____
   _____

10. What are acceptable or important reasons (ex. procreation, making up after a fight, self-validation, etc.) for people to engage in sexual activity? Rank your top 3 reasons in order of importance to you. _____
_____
_____
_____

11. What criteria do you use to determine if you or someone else is ready to engage in sexual activity? _____
_____

12. Whose responsibility is birth control or family planning? _____
_____

13. What was the attitude toward pornography in the home you grew up in? What are your current attitudes about pornography? _____
_____
_____

14. What are your attitudes regarding masturbation? For men? For women? _____
_____
_____

How do your religious values or beliefs inform your ideas or attitudes about sex?
_____
_____
_____

15. What are other experiences you have had that influence your beliefs about sex?
_____
_____
_____

WHERE ARE YOU GOING?
MAPPING YOUR RELATIONSHIP INTIMACIES

Evaluate the different types of intimacies in your current relationship and invite your partner to map their interpretation of the intimacies as well. You may also map a past relationship or an ideal future relationship. Draw a circle to represent a pie chart. Divide the circle into seven different sections, or slices, according to how much attention or focus is given to different "intimacies" of the relationship. These different aspects of intimacy are:

- ✓ *Sexual intimacy* (closeness felt based on sexual experiences)
- ✓ *Emotional intimacy* (closeness based on understanding and connecting emotionally; understanding each other's emotions)
- ✓ *Physical intimacy* (physical closeness that is non-sexual like holding hands, sitting close, physical touch)
- ✓ *Psychological intimacy* (closeness based on similar thinking; mentally connecting)
- ✓ *Recreational intimacy* (closeness based on participating in activities together; supporting each other in activities)
- ✓ *Spiritual intimacy* (closeness based on common spiritual or religious beliefs; deep core values)
- ✓ *Intellectual intimacy* (closeness based on stimulating exchange of ideas; similar levels of intelligence; talking on the same "level")

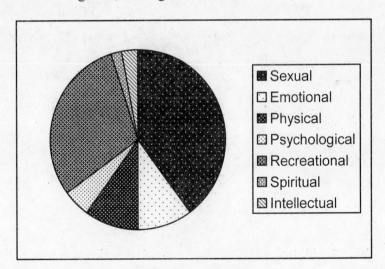

Analyze your breakdown of the relationship you mapped out.

- ➢ Which intimacies seem strongest?
- ➢ Which intimacies could be strengthened? What would you do to strengthen them?
- ➢ Are there any components of the relationship that are overshadowed or ignored?
- ➢ Is sexual intimacy too much or too little of a focus? Does it seem balanced with the other intimacies?
- ➢ How do the other intimacies relate or connect to sexual intimacy?
- ➢ How can a singular focus on sexual intimacy crowd out other important relationship intimacies?

Share with your partner your interpretation and impressions and have them do the same with you.

> How similar or different are the various intimacies experienced in the relationship by each of you?
> Is each of you satisfied with the current portrayal of the intimacies?

DEFINING INTIMACY

How do you define intimacy? To answer this, you may think of things you have noticed in a past or current relationship, or think of what you anticipate in a future relationship. You may also have a partner respond to these questions in writing, and then discuss and compare answers as a couple.

- How do you define intimacy?
- What is the difference between physical intimacy and emotional intimacy?
- What do you understand to be the connection, or link, between physical intimacy and emotional intimacy? What have you noticed about this connection in your relationship, either current or past?

WHAT'S YOUR COMFORT LEVEL?
A HOMEWORK ACTIVITY

How comfortable are you discussing sex with your partner?

Work on a level of comfort in discussing sex as a couple by picking an informative book on sex to read aloud together. The book (or textbook) should explain and portray male and female anatomy, anatomical terms, physiology and physiological responses of the body, as well as portray the sexual response cycle. Build a knowledge platform and forum to talk about sex by setting aside time where you and your partner can read this book together, taking turns reading aloud to each other, allowing time to pause for discussion or to ask questions. It may take several reading sessions to complete this exercise dependent on the length of the book. The goal should be to provide opportunities for basic learning and appreciation of your own and your partner's body and each other's physical ability to respond sexually. This will also help work toward a level of comfort in discussing sex and asking questions of each other.

How comfortable are you with your sexual knowledge?

Become familiar with male and female anatomy, as well as the sexual response cycle, using a human sexuality textbook or other book on human sexuality, or notes from a documentary film on human sexuality. Using books or other pictures (anatomy drawings or overheads) as a reference, teach a 20-30 minute mini-lecture to a small group of friends, roommates, or relatives. Be sure to educate about anatomy as well as what responses occur for men and women during each phase of the sexual response cycle. You can also allow time for discussion or questions from the group.

GIVING AND RECEIVING FEEDBACK ABOUT PHYSICAL TOUCH
A HOMEWORK ACTIVITY

Invite your partner or friend to do this exercise with you. One person will start out as the Giver, and one as the Receiver. The Giver will give a hand, foot, or shoulder massage to the Receiver for 10 minutes. The Receiver should give instruction and feedback to the Giver on how she or he would like to be massaged, whether the action needs to be harder, softer, faster, slower, or what areas may need special attention or extra massage. At the end of 10 minutes roles are switched with the Receiver of the massage now being the Giver. Again, feedback and instruction should be given by the Receiver. At the end of 10 minutes process your experience together.

- ✓ What was it like to give feedback to someone regarding what you liked or didn't like about their touch?
- ✓ What emotions came up during each of your roles as Giver or Receiver (fear, anxiety, frustration, relaxed, etc.)?
- ✓ As a Receiver, what was it like for you to give feedback or requests for change? Did you feel you could make requests about what you preferred in terms of touch?
- ✓ As a Giver, did you feel you understood your partner's requests? Did you ask questions for clarification? Why or why not?
- ✓ As a Giver, were you glad to adjust your touch or did you proceed as you had been? Were you open to feedback, or did you find yourself giving the type of massage you know *you* would enjoy?
- ✓ How would you relate this exercise to sexual communication in a relationship?

If doing this exercise with your partner, also discuss the following:

- ➢ How does this communication exercise about touch relate to your sexual relationship?
- ➢ How can you make aspects of listening, giving feedback, and making requests, a workable part of your sexual relationship?

Chapter 5

FINANCIALLY EVER AFTER

WHAT ARE YOUR FINANCIAL VALUES?

We tend to spend money on things that reflect our values. What are your core values that may influence your financial decisions? Fill out the "Financial Values Inventory" (p. 51) ranking your top 10 answers. Contemplate the following questions:

- What do you value?
- Based on this, what do you tend to spend your money on?
- Are there things you spend money on that you do not see as reflective of your values?
- What types of realizations or connections did you make between what you value and what you buy or spend money on?

You may also fill out the "Financial Values Inventory" by predicting what your partner's responses may be. Have your partner fill out their own survey assessing which ten items are most important to them and also what they would predict for your answers. Compare and share your answers and predictions as a partnership.

- Are you aware of what each other value?
- How did each of you do in forecasting what your partner values? Based on these values, what would each of you predict your partner may spend money on, or invest money in?
- Are there fairly major expenses either of you have that do not reflect what is on your Top 10 list?
- Which responses or predictions surprised you?

FINANCIAL VALUES INVENTORY

Pick the top 10 items from the list that are of most value to you and rank them. You may also predict which 10 items you think would be of most value to your partner and rank them in order of importance.

You	Your Partner	
___	___	A secure and comfortable retirement
___	___	A sense of equality in relationships
___	___	Emotional and sexual intimacy
___	___	A sense of accomplishment in life
___	___	A sense of independence and self-reliance
___	___	A meaningful love relationship
___	___	Financial security for the family
___	___	Happiness or contentedness
___	___	A meaningful personal spirituality
___	___	Feelings of self-confidence
___	___	Social recognition and community status
___	___	A fulfilling marriage
___	___	Meaningful purpose in life
___	___	Helping the poor, sick and disadvantaged
___	___	A sense of family togetherness and happy children
___	___	Learning, gaining knowledge continually
___	___	Honesty and personal integrity
___	___	Good health and physical fitness
___	___	Close relationships with extended family
___	___	Traveling and quality vacations
___	___	Companionship, spending time together as a couple
___	___	Success in a job or career
___	___	Freedom to live life as you choose
___	___	New experiences and adventures
___	___	Being outdoors, away from city life
___	___	Satisfying friendships, liking people and being liked
___	___	Living in the city, access to restaurants and entertainment
___	___	Time alone, being by yourself
___	___	Having nice things, such as cars, boats, furniture
___	___	Emotional security, freedom from excessive stress

From Poduska, B. E. (1995). *For Love & Money.* Pacific Grove, CA: Deseret Book Company and Brooks/Cole, p. 27.

10 FINANCIAL PRINCIPLES

Review the "10 Financial Principles" information sheet (p. 53) individually or together with your partner.

- Which principles resonate as principles either you or your partner practice in your relationship?
- Which principles are new to you or even surprising?
- Which principles do you think describe you or your relationship (even the negative interpretation)?
- Based on these principles, what changes do you feel you need to make, either individually or as a couple?

10 FINANCIAL PRINCIPLES

1. **Financial problems are usually behavior problems rather than money problems.**
    Having more money doesn't mean you will manage it any better.

2. **If you continue doing what you have been doing, you will continue getting what you have been getting.**
    If you keep spending instead of saving, you will keep living with no savings.

3. **Nothing (no thing) is worth risking the relationship for.**
    You can destroy a marriage by loving "things" more than loving each other.

4. **Money spent on things you value usually leads to a feeling of satisfaction and accomplishment. Money spent on things you don't value usually leads to a feeling of frustration and futility.**
    Focus your resources on goals that are connected to your values.

5. **We know the price of everything and the value of nothing.**
    We often feel pressure to buy things we think we *should* want instead of buying what we *really* want.

6. **You can never get enough of what you don't need, because what you don't need can never satisfy you.**
    Purchasing "things" to make up for intangibles like love, attention, or time will not satisfy, because the "things" are not what you really need.

7. **Financial freedom is more often the result of decreased spending than of increased income.**
    Financial freedom is having financial alternatives instead of financial ultimatums.

8. **Be grateful for what you have.**
    The quest for more "stuff" will never end. Be content with what you have.

9. **The best things in life are free.**
    Love, humor, and friendship do not cost a thing.

10. **The value of an individual should never be equated with his or her net worth.**
    The worth of an individual cannot be put to a dollar sign.

The ten financial principles are from Poduska, B. E. (1995). *For Love & Money.* Pacific Grove, CA: Deseret Book Company and Brooks/Cole, p. 277.

YOU BOUGHT *WHAT*?
ATTITUDES AND RULES THAT INFLUENCE OUR FINANCES

What is your attitude toward money? What did you learn from your family of origin regarding money and finances? What financial "rules," either explicit or implicit, were followed in your family? Fill out your own copy of the "Deciphering Our Attitudes Toward Money" inventory (pgs. 56-58) and contemplate or write down answers to the following:

- What are some "meanings" money had in the family you grew up in?
- Was money seen as a resource, or as a weapon in the house you grew up in?
- Was money ever used as a substitute for intangible things like love, attention, time, or to say "I'm sorry?"
- How do you feel about the "knowledge" you obtained about money?
- Are there any messages or meanings that you don't like? What would you like to change?

Have your partner fill out the "Deciphering Our Attitudes Toward Money" inventory. As a couple, discuss and compare the message you each learned in your family of origin about money.

- What messages regarding money did your partner learn in their family of origin?
- What are similar messages you both received growing up about money and finances?
- For which questions are your experiences or what you learned vastly different?
- Does understanding the "meaning" behind the money for each partner, help explain any financial disagreements you have had?

Take time to ponder, or for you and your partner to discuss, the experiences you had in your family of origin regarding money and the feelings and attitudes you have learned to associate with it. Think of these messages you received about money in terms of spoken and unspoken rules (or ways of behaving) that you internalized. For example, if your parents bought everything with cash or check, the rule you learned could be "Never go into debt." This rule could be implied (unspoken) or actually verbally expressed in your family. Work to come up with a list of financial rules you have learned in your family. Your partner should also come up with their own list. List these rules on the "What Were Your Family's Financial Rules?" worksheet (p. 59), deciding if the rule learned was something learned by actual verbal expression, or if it was taught nonverbally by example.

Next, look at your own list of rules and answer the following:

- ✓ What are the rules about money that you learned in your family of origin?
- ✓ Did they tend to be taught explicitly or implicitly?
- ✓ How do you feel about these rules?
- ✓ Which rules do you feel work for you now, and will probably work in the future?
- ✓ Which rules do you want to ignore or stop following?
- ✓ What are some new financial rules you want to adopt or practice?

Compare your list of rules with your partner's list.

- ✓ Did you have any rules in common?
- ✓ Are there rules explicitly taught in one partner's family but were implicit in the other partner's family?
- ✓ Which rules contradict each other (ex. men should manage household finances vs. women should manage household finances)?
- ✓ In general how do you and your partner feel about the rules you've learned about money?
- ✓ As a couple, which rules do you want to keep for your relationship? Which rules need to change?
- ✓ Do you feel you are on the "same page" financially with your partner? If no, what would it take to get there?
- ✓ What new rules would you like to institute or follow as a couple?

WHERE DID YOU GET THAT ATTITUDE?
DECIPHERING OUR ATTITUDES TOWARD MONEY
AN INVENTORY

What is your attitude toward money? Answer the following questions to determine what you learned in your family of origin about the meaning behind money and resources. You may also have a partner fill out their own inventory.

1. Do you think the family you grew up in was materialistic? In what ways? _____
   _____
   _____
   _____

2. Could you ask for financial support from you parents or family members? What usually happened if a family member asked for financial support? How did others react? Did family members have to ask, or was financial support just expected or given? How were children in your family financially supported? How much, how often, and for what? Did age (over or under 18) or stage of life matter (living at home vs. living outside the home)? _____
   _____
   _____
   _____
   _____
   _____
   _____
   _____
   _____
   _____

3. Was affection ever expressed between you and your parents in terms of money or gifts? What did this look like? How often did it occur? _____
   _____
   _____

4. Was affection ever expressed between your parents in terms of money or gifts? What did this look like? How often did it occur?
   _____
   _____

5. Did your parents express approval or disapproval with money or gifts? What did this look like? How often did it occur? _____
   _____
   _____

6. How did your family evaluate success? How did money, educational degrees, home ownership, social status, or possessions fit into this definition? In what other ways was success defined? _____
_____
_____
_____

7. How did your parents or the family you grew up in feel about debt? _____
_____
_____

8. How were the responsibilities of managing finances divided up or handled in the family you grew up in? Who did what? _____
_____
_____
_____

9. Did the family you grew up in talk about finances? Who participated in those conversations? Adults? Teens? Children? _____
_____
_____

10. In which socioeconomic class (middle class, upper class, etc.) do you feel the family you grew up in belonged? How did you know? How did you feel about your family's status? Did your family's social class change at all over the years?
_____
_____
_____
_____

11. In the home you grew up in, who worked outside the home? Why? _____
_____
_____
_____

12. In the home you grew up in, what was the attitude toward saving or investing? What is your current attitude about saving money? What has influenced that?
_____
_____
_____

13. In the home you grew up in was there a steady monthly income? How did that feel to you? _____
_____
_____

14. Are you someone who has the money for something before you buy it, or do you usually buy things first and figure out how to pay later? _____
_____

15. How were credit cards used in the family you grew up in? Do you have credit cards? How many? What kind? How do you use them? _____
_____
_____
_____
_____

16. Do you usually have money left over at the end of your pay period or do you tend to live "paycheck-to-paycheck"? Why? _____
_____
_____

17. Do you have a checking account or checkbook? Who maintains it? Why? _____
_____
_____

18. Who is responsible for the actual payment of your monthly bills (rent, mortgage, car payment, credit cards, utilities, etc.)? Why? _____
_____
_____

19. If it were possible to save $200 a month, what would you do with that money? Please rank the following answers, using a 1 for what you would be most likely to do with the $200, and 10 for the least likely.

    _____   save it for a rainy day
    _____   save it to buy something for cash rather than credit
    _____   invest it
    _____   use it for recreation (games, movies, sports)
    _____   use it for payment for a new car or other large purchase
    _____   use it for travel and adventure
    _____   use it to buy a home or property
    _____   use it to improve your present living space
    _____   divide it in half with a partner or a friend and let each person spend it as he or she chooses
    _____   use it for a new wardrobe, eating out, or entertainment

WHAT WERE YOUR FAMILY'S FINANCIAL RULES?

Explicit Family Rules

Financial rules actually talked about or verbally expressed in your family.

Examples: "Don't waste money on junk food."
"You only live once—go ahead and buy it."

1. _____
2. _____
3. _____
4. _____
5. _____
6. _____
7. _____
8. _____
9. _____
10. _____

Implicit Family Rules

Financial rules taught by example, through nonverbal communication or intuition (you just "knew").

Examples: "Men are better managing money than women."
"Whoever makes the money decides how to spend it."

1. _____
2. _____
3. _____
4. _____
5. _____
6. _____
7. _____
8. _____
9. _____
10. _____

HOW SATISFIED ARE YOU WITH YOUR FINANCIAL TASKS?

Are you and your partner satisfied with the financial tasks each of you are doing? Fill out the "Managerial Task Satisfaction Scale" (p. 61), marking your responses to these items. Also indicate how you feel your partner would respond in terms of their own satisfaction with the current arrangements. Have your partner also fill out their own copy of the scale, indicating their own satisfaction and how they perceive your satisfaction with the allocation of financial tasks.

As a couple, discuss and compare answers, both actual and predicted.

- ✓ What are task arrangements one partner is unhappy or happy with, but the other partner perceives the situation differently?
- ✓ Which tasks are either of you happy to do?
- ✓ Which tasks do neither of you enjoy doing?
- ✓ Based on your current arrangement, which tasks can be reshuffled or approached differently in order to make both you and your partner more satisfied with the arrangement?
- ✓ If one partner has a skill set in one area (like math) should that person automatically take on all things number-related (taxes, writing checks, paying bills)?
- ✓ What skills does each of you have that you could teach the other?
- ✓ Is it valuable to learn new financial managerial skills, or should each partner stick with what she or he knows best?
- ✓ What are some areas where neither of you is knowledgeable? What are resources to seek out or skills to be developed? How will you accomplish this as a couple?

HOW SATISFIED ARE YOU WITH YOUR FINANCIAL TASKS?
MANAGERIAL TASK SATISFACTION SCALE

Who performs the following financial management tasks in your home? Are you satisfied with how responsibilities are divided? Rate yourself on a scale of 1-5 by marking the accurate number. Using a different symbol or color, also indicate how much you feel your partner is satisfied with the arrangement.

	Completely Happy				Completely Unhappy
1. Shopping for, buying groceries	1	2	3	4	5
2. Obtaining maintenance, service, and repairs for the car(s)	1	2	3	4	5
3. Shopping for, selecting, and purchasing new or used cars	1	2	3	4	5
4. Studying, deciding on, and investing in property, stocks, and bonds	1	2	3	4	5
5. Studying, deciding on, and purchasing life, hospital, and medical insurance	1	2	3	4	5
6. Studying, deciding on, and purchasing car, fire, liability, and other property insurance	1	2	3	4	5
7. Figuring annual federal and state income taxes	1	2	3	4	5
8. Maintaining records of income and expenses	1	2	3	4	5
9. Preparing monthly or annual budget	1	2	3	4	5
10. Paying bills	1	2	3	4	5
11. Signing checks and making deposits	1	2	3	4	5
12. Earning money through employment	1	2	3	4	5
13. Assuming responsibility for family estate, will, and related matters	1	2	3	4	5
14. Obtaining medical and dental care	1	2	3	4	5
15. Managing family time commitments	1	2	3	4	5
16. Deciding on and performing *inside* chores	1	2	3	4	5
17. Deciding on and performing *outside* chores	1	2	3	4	5

From Poduska, B. E. (1995). *For Love & Money.* Pacific Grove, CA: Deseret Book Company and Brooks/Cole, p. 49.

HOW'S YOUR BUDGET?

How does your income compare to your expenses? Fill out the "Budgeting" worksheet (pgs. 63-64) independently or with your partner, determining income, fixed expenses, and variable expenses. These values can be based on what you actually earn or pay, or what you perceive you will earn and pay for the next month. After filling in all expenses, look at the list again and determine which of the expenses are actual "needs" and which debts would fit more in a category of "wants." Be sure to deconstruct larger payments that may be in fixed expenses. For example, your car payment may be a fixed expense, but does the payment reflect what you *need* (a used compact size car to travel to and from work) or what you *want* (a new 4-wheel drive SUV, with heated seats, XM satellite radio, and flip-down DVD player and screen).

- How does your budget look? Are you able to allocate for necessary expenses?
- As you, or you and your partner, evaluate your outgoing expenditures in terms of needs and wants, what does this change about your debt to income ratio?
- Are there things you are spending money on either independently or as a couple that you wish you weren't?
- What things were thought of as "needs" at the time you bought them, but you now recognize them as "wants?" What can you do to adjust your expenses?
- How do you, or you and your partner, feel about your monthly "bottom line?" If you are unhappy with it, what is in your control that can change?

BUDGETING

Record your income and expenses as an individual or as a couple who has pooled their resources. For expenses, estimate the amounts spent monthly in each category. If there are categories that you don't spend monthly on (every 6 months, or once a year), divide the payment in 12 to estimate how much you would need to allocate per month.

Income

Partner 1 Salary/wages _____
- Less: Withholding taxes _____
- Retirement _____
- City or local taxes _____
- Any other withholding _____

NET INCOME _____

Partner 2 Salary/wages _____
- Less: Withholding taxes _____
- Retirement _____
- City or local taxes _____
- Any other withholding _____

NET INCOME _____

Other income _____
- Less: Any other withholding (e.g. alimony, child support) _____

TOTAL NET INCOME _____
(sum)

Expenses

Fixed Expenses
- Rent or mortgage _____
- Property Taxes _____
- Insurance Premiums _____
- Car license/registration _____
- Debt repayment (ex. car, loans, creditors, credit cards) _____
- Other _____

TOTAL FIXED EXPENSES _____
(sum)

Variable Expenses

    Food/Groceries   _____
    Utilities (e.g. gas, electric, water, garbage)   _____
    Phone (landline and cell)   _____
    Gasoline   _____
    Other transportation (e.g. bus, train)   _____
    Vehicle maintenance   _____
    Furnishings/equipment   _____
    Household operations   _____
    Household repairs   _____
    Education   _____
    Clothing   _____
    Childcare   _____
    Toiletries/household products   _____
    Pets (e.g. food, care, veterinarian visits)   _____
    TV/cable/dish/internet   _____
    Entertainment   _____
    Recreation   _____
    Memberships (e.g. gym, country club, pool)   _____
    Eating out (including snacks at gas stations)   _____
    Magazines/newspapers   _____
    Haircuts   _____
    Hobbies   _____
    Personal allowances   _____
    Out of pocket medical co-pay   _____
    Gifts/donations (e.g. Christmas, birthday, special occasions)_____
    Other _____   _____

    TOTAL VARIABLE EXPENSES   _____
                                                                        (sum)

TOTAL FIXED EXPENSES   _____ (+)
TOTAL VARIABLE EXPENSES   _____ (=)
TOTAL EXPENSES   _____

TOTAL NET INCOME   _____ (-)
TOTAL EXPENSES   _____ (=)
INCOME TO DEBT DIFFERENCE   _____

DOING AWAY WITH DEBT

How can you eliminate creditor and credit card debt? Chart your credit card and other cumulative debt on the "Doing Away With Debt" worksheet (p. 67). Keep in mind as you are paying down cumulative debt from credit cards that you cannot keep using the cards.

DOING AWAY WITH DEBT: SCHEDULING YOUR DEBT PAYMENTS

Month	Credit Card 1	Credit Card 2	Credit Card 3	Credit Card 4	Car Loan
October	100	85	50	75	350
November	100	85	50	75	350
December	100	85	50	75	350
January	100	85	50	75	350
February		**185**	50	75	350
March		**185**	50	75	350
April		**185**	50	75	350
May			50	75	350
June			**235**	75	350
July			**235**	75	350
August				**310**	350
September				**310**	350
October					**660**

Mapping out a schedule to pay off debt and creditors can help you reduce or eliminate debt. Use this calendar as a reference and the next worksheet page to map out your own debt repayment schedule. In the very left column, write the names of the months of the year, beginning with the upcoming month. You may extend the length of the calendar to reflect any number of months or years. On the top of the other columns, write the names of the creditors or credit cards that need to be paid off. You may want to first pay off debts that have the lowest amount to be paid off or that have the highest interest rate. List in the first creditor column what the monthly payment would be to that creditor until the debt is repaid, as shown above (ex: Credit Card 1, $100 per month). In the next column, under the name of the second creditor (and third, fourth, fifth, etc.), list the minimum payment due each month or what you can pay to these additional creditors. After you have completely repaid the first creditor, add the amount that you were paying for that monthly payment to your payment for the next creditor. In the table above, the individual or couple finished making monthly payments to Credit Card 1 in January. The $100 usually used for that payment was then added to the $85 payment for Credit Card 2, creating a new monthly payment of $185 going to Credit Card 2. Continue this process until all loans and debts are repaid. Keep in mind you must stop using the creditor or credit card for this debt repayment schedule to work (don't keep using the credit cards and ringing up new charges).

Adapted from Ashton, M. J. (2006). *One For the Money: Guide to Family Finance.* Salt Lake City, UT: The Church of Jesus Christ of Latter-day Saints, p.5.

DOING AWAY WITH DEBT WORKSHEET

Month	Debt #1 ___ —	Debt #2 ___ —	Debt #3 ___ —	Debt #4 ___ —	Debt #5 ___ —	Debt #6 ___ —	Debt #7 ___ —

Use this worksheet to create your own debt repayment calendar. Add rows as necessary to extend the time needed to pay off all debts. Add columns as necessary to track your repayments to all creditors.

FINANCIAL INTERVIEW
A HOMEWORK ACTIVITY

It is essential that you and your partner implement a plan to successfully manage your finances as a couple. Oftentimes we learn what will and won't work for us by seeing how other couples manage their finances.

Pick one or two couples or family members in a long-term relationship who you can interview, preferably together, either in person or on the phone. The goal is to learn about how other couples manage their finances and divide up financial managerial duties. Listed below are some sample questions you can use to conduct the interview. Ask any follow up or clarifying questions as appropriate. This activity may also be done without a partner, focusing on what you think will or won't work in your own future relationship.

Possible questions to ask:

- ➢ Do you believe that all your assets belong to both of you and that all your debts are your joint responsibility?
- ➢ Do you share equally in financial decisions? Why or why not?
- ➢ Do you use a budget? Why or why not?
- ➢ How do you feel about each partner having a personal allowance for which she or he is not accountable to the other?
- ➢ How do you feel about buying on credit? Small items? Large items?
- ➢ What did you do about the debts each of you brought into the relationship?
- ➢ Who is the bill payer? Check writer? Bookkeeper? Investor? How is this working for you?

Additional ideas for questions can be taken from the "Managerial Tasks Satisfaction Scale" questionnaire (p. 61).

After the interview, discuss or write down your impressions of the interview.

- ✓ What did you like or not like about what the interviewed couple did to manage their finances?
- ✓ Are there any practices you would accept or adopt?
- ✓ What do you feel wouldn't work for you as a couple?
- ✓ Where do you go from here in terms of mapping your own financial future as a couple?

Chapter 6

MARRIAGE ENRICHMENT

NURTURING A RELATIONSHIP OVER TIME

What does it take for a relationship to stay together over time? How can a couple survive all of the trials and transitions that life brings and still come out on top? Oftentimes it seems that an enduring relationship is attributed to good luck, or somehow it "just happens." The reality is that a lasting relationship takes work, conscious deliberate effort, forgiveness, and a positive outlook. It means reaching out and including someone else in your life and your decisions. It involves being respectful and nice to your partner even when it is tough.

Sometimes we think of marriage enrichment as something that is done after a couple has been together a long time, or something that occurs at retirement age. This is not true. Marriage enrichment is a continual process, something that is contributed to steadily by both you and your partner.

Listed here are several ideas couples can utilize to continue to promote mutual learning and respect for each other and their relationships. By no means is this an exhaustive list, but rather something to help couples get started on their own road of marriage enrichment. As you contemplate what you need to enrich your own relationship, keep in mind it is not so much what you do—it is how you do it, and why.

SUGGESTIONS FOR NURTURING A RELATIONSHIP OVER TIME

1. Read a book together as a couple that addresses healthy relationships or how to have a healthy marriage. One book suggestion is *Why Marriages Succeed or Fail…and How You Can Make Yours Last* by John Gottman. The book outlines Gottman's well-known research in a user-friendly way and even contains self assessments and quizzes to evaluate the presence of unhealthy relationship interactions.

2. Be sure to plan time to be together as a couple for "dates" even if it's just dinner or a time to sit down for meaningful conversation. As life gets busy, work is hectic, and children are a priority, time together does not happen spontaneously. There is nothing wrong with planning some "we" time into the schedule.

3. Schedule a "marriage check-up" with a professional counselor, clergyman, or a trusted spiritual leader. Talking to an objective third party is often helpful and it does not have to mean your marriage is in trouble. Often these professionals have access to assessments or materials that can help point to areas of the marriage to work on, as well as things that are going well. Encouragement from a supportive professional can give you the boost you need.

4. Serve and appreciate each other with "gifts" from the heart. Each of you write down ten things or "gifts" that communicate love to you, and you would like to receive from your partner. Anything can go on the list, but small and simple are best. Some examples are:

    - A hug after work
    - Vacuuming the living room
    - A 5 minute shoulder massage
    - Planning a night out for the two of us
    - A note of affection left on the bathroom mirror

    After each of you have your lists done, exchange them. For the next seven days, each person picks one thing off of their partner's list to do, with a new item being picked each day. Each partner does their act of service unannounced and it is the receiving partner's job to notice the act and express appreciation.

5. Go to a local library together and research "marriage enrichment." Locate and read any interesting research articles or books that come up.

6. Be creative in how you spend the time together that you do have. Don't just watch TV or a movie, think of other things you enjoy doing together.

7. Make a list with your partner of inexpensive date ideas—things you can do together that do not cost money or are under $5. Next time you have a date night, or maybe just 30 minutes, pull out the list and pick something to do. For example:

    - Go to a store and try on ridiculous clothes or shoes that are *not* you. Take pictures.
    - Go to a grocery store with $3 for each partner to spend on ingredients for dinner. Go your separate ways to spend your money and then meet up in 20 minutes (each person doesn't know what the other bought). Be creative in assembling and planning your dinner for that night.
    - Take a walk in the rain.

8. Take an extended car trip together that is at least an hour of driving. Turn off the radio or CD player and just talk while you drive.

9. What new hobbies can you and your partner do together? What have you always wanted to do together, but you just haven't yet? Or, how can you be more supportive or involved in the hobbies your partner currently has?

10. Interview couples you know who are 5, 10, or 20 years further down the road of life than you are. These can be couples who are facing transitions (having young children, raising teenagers, having children leave home, facing retirement) that you are not facing now, but will soon. What advice would this couple give you? What do they have to share?

11. In what capacity can you and your spouse serve in the community? Find ways to share your life experience and who you are, as well as learn from others. Not only does serving others feel good, but it can help bring a new purpose to your partnership.

12. Laugh at yourself and your mistakes. Try your best to do better as a partner and as a couple. Neither of you is perfect. A good marriage is part of a continual learning and growing process.